# YOUR KNOWLEDGE HAS VALUE

- We will publish your bachelor's and
  master's thesis, essays and papers

- Your own eBook and book -
  sold worldwide in all relevant shops

- Earn money with each sale

Upload your text at www.GRIN.com
and publish for free

# The Above and Below Ground Biomass and Carbon Stock Estimation Options for Trees and Forest Resources in Mago National Park, Southern Ethiopia

**Bibliographic information published by the German National Library:**

The German National Library lists this publication in the National Bibliography; detailed bibliographic data are available on the Internet at http://dnb.dnb.de.

ISBN: 9783346867339
This book is also available as an ebook.



Print and binding: Books on Demand GmbH, Norderstedt, Germany
Printed on acid-free paper from responsible sources.

The present work has been carefully prepared. Nevertheless, authors and publishers do not incur liability for the correctness of information, notes, links and advice as well as any printing errors.

GRIN web shop: https://www.grin.com/document/1353485

Review on the above and below ground Biomass and Carbon Stock Estimation options for Trees and Forest Resources: Within and Around Mago National Park, Southern Region of Ethiopia.

Wondo Genet College of Forestry and Natural Resources, Hawassa University.P.O. Box.128, Shashemene.

## *Abstract*

*Forests play an essential role as source and sink in global carbon cycle. Development and other human induced activities have led to degradation of forest land, and ultimately, it results in loss of biodiversity and increases concentration of $CO_2$ in atmospheres. Therefore, there is urgent need to estimate regional and national level carbon stock for making forest-based policies and strategies for mitigation of $CO_2$. Variable and periodic information is available on biomass and carbon stock of Gambo Natural Forest. This review presents a systematic review on the Biomass Carbon Stock Estimation Options for Trees and Forests Resources: Within and Around Mago National Park, Southern Region of Ethiopia. In order to review and gain information from existing research on types of forest, basal area, biomass/carbon stock, sequestration pool in different forest ecosystems of Mago National Park, a literature search was carried out during January 2022 to October 2022 using Web of Science, Google Scholar Citation, Research Gate,IPCC, United Nations Framework Convention on Climate Change (UNFCCC) and REDD+ report, offline journals, book chapters, , government scientific reports, Forest Survey data, Botanical Survey data, as well as reports published by Ministry of Environment, Forest and Climate Change. The review is limited to in and around Mago National Park forest ecosystem in relation to above and below biomass and carbon stock estimations. Formulating allometric equations for all woody plants in Ethiopia is quite desirable for accurately quantifying the biomass and carbon stock of forests to achieve accurate national and international reporting of carbon dioxide emission inventories. Depending on this in Ethiopia several study undertaken regarding allometric equation for biomass estimation, but it is not as forest resource of the country. Most of the study undertaken was depend on semi-destructive method which is not accurate as destructive measurement.*

*Key words: Allometric equation, Biomass, Mago National Park, Carbon Stock, Forest Resources*

# 1. INTRODUCTION

## 1.1. Background

Forest ecosystems act as source and sink of atmospheric carbon dioxide ($CO_2$) and are one of the most faithful options for carbon sequestration and play a crucial role in regulating global carbon cycle. Local, regional, and national carbon inventories of source and sinks of carbon are indispensable to assess the prospective role of various carbon sequestration pools for reducing atmospheric $CO_2$ accumulation, and therefore it is a pioneer step for preventing global warming. The studies also important for developing of systems/markets for national and international carbon credit/emission trading as well as in reducing emission from deforestation and forest degradation (REDD+) programs in developing countries (Han et al. 2007; NEFA 2002; Kale et al. 2004).

Growing concerns about climate change resulting from increased concentration of greenhouse gases in the atmosphere have stimulated discussions about the importance and potential of forests and Agroforestry for carbon sequestration. Due to anthropogenic emissions, the concentration of the major greenhouse gas, carbon dioxide ($CO_2$), has reached 400 parts per million (ppm) for the first time in record history, according to report from the Mauna Loa Observatory in Hawaii in 2015 (MLOH, 2015). The globally averaged combined land and ocean surface temperature data show a warming of 0.85 [0.65 to 1.06] °C over the period 1880 to 2012. The total increase between the middling of the 1850–1900 period and the 2003–2012 period is 0.78 [0.72 to 0.85] °C. Regional temperatures may increase even by 1 to 5 °C, if the current atmospheric CO2 concentration is doubled (IPCC, 2014).

To reduce the rising levels of greenhouse gases, in particular CO2, afforestation and reforestation systems have been encouraged as means to sequester CO2 in biomass and soil, an idea formally recognized by the Kyoto Protocol. The Kyoto Protocol agrees for the opportunity to offset CO2 emissions through cooperation between developed and developing nations to undertaking into reforestation or afforestation projects (UNFCCC, 1997).

Forest ecosystem is a major constituent of the carbon reserves and it plays an important role in regulating global climate change through process of carbon sequestration (Addi et al., 2019; Tadesse et al., 2019). Tropical forest is a main constituent of terrestrial carbon cycle and it has a great potential for carbon sequestration, accounting for 26% carbon pool in aboveground biomass and soils (Brassard et al., 2009).

Agroforestry is also recognized as one of the greenhouse gas lessening strategies under the Kyoto Protocol, which was adopted in 1997. As a result, the sequestration potential of agroforestry systems has gained the attention of many countries throughout the world, prompting them to concentrate on it. Several authors have suggested that trees in agroforestry practices absorb and store larger quantities of atmospheric carbon dioxide ($CO_2$) than do the herbaceous seasonal or annual crops and pastures (Nair, P.R. et al., 2009 ; Sharrow, S.and Ismail, S., 2004). This is because incorporation of perennial trees and shrubs in croplands and pastures would result in higher Carbon sequestration both in biomass and the soil (Nair, P.R. et al., 2009 ; Haile, S.G. et al., 2008). Therefore, improving C stock in farmlands by introducing AF practices could be a potential option to mitigate climate change impacts (Gebremeskel, D. et al., 2021). Biomass assessment of tropical forests and Agroforestry are crucial for appreciative the role of terrestrial ecosystems to the carbon cycle and climate change mitigation (Ali et al., 2015; Ancelm et al., 2016).

Reliable estimation of forest biomass with a sufficient accuracy is crucial to measure the variations of Carbon Stored in the forest (Ketterings et al., 2001; Chave et al., 2004). It is a fundamental aspect of studies of Carbon stocks and the effect of deforestation and Carbon sequestration on the globe Carbon balance. Measuring biomass in local, regional and global scales is critical for estimating global carbon storage and assessing ecosystem response to climate change and anthropogenic disturbances (Ni-Meister et al., 2010).

Accurate estimations of biomass in tropical forests are deficient in many areas, and this is due to the lack of appropriate allometric models for predicting a biomass in species-rich tropical ecosystems and such paucity of information makes estimation of the value of these species as carbon sinks difficult (Chave et al., 2005). There is a lot of uncertainty in the amount and spatial variations of above-ground biomass in Africa, partly because very few allometric equations are available (Fayolle et al., 2013).

The use of existing generalized biomass equations across wider ecological zones can lead to a bias and error in estimating biomass for particular species and sites (Henry et al., 2011) because, there are differences among species in wood specific gravity, tree sizes, growth stages, and also since some geographic areas have not been covered by the existing general allometric equations is that it cannot be applicable for these types of geographic areas which are available for similar species (Navar et al., 2002).

In addition, the accuracy of biomass estimations can be affected by several factors such as climate, topography, soil fertility, water supply, wood density, distribution of tree species, tree functional types and forest disturbances have impact on forest variability (Fearnside, 1997; Slik et al., 2008). For a determined tree species, tree mass is influenced by the size of the tree, its architecture, (form), and health (e.g. hollow trees), (Fearnside, 1997).

Allometric equations are a basic tool for non-destructive estimation of biomass in woody vegetation. These equations express tree biomass as a function of easy-to-measure parameters such as diameter, height, volume or wood density, or a combination of thereof (FAO, 2012; Brown, 2002; Chave et al., 2014). Equations generated from a small sample of trees are then used to estimate biomass at plot level and landscape scales. It's also the convenient and common method to estimate the biomass of a forest or stand. Biomass estimation through allometric equation for forest and Agroforestry are very vital for mitigation of climate change and sustainable management of the two systems (Wang, 2006).Thus, the overall objective of this review is to identify what is known and what is not known (research gap) on the following specific objectives.(i) To develop allometric equation for biomass carbon stock estimation carbon for selected important tree species of Mago National Park forest ecosystem.(ii)To determine uncertainties associated with using species specific, site specific and generic biomass models for forest biomass estimation.(iii) To estimate the biomass carbon stock in harvested wood products for selected tree species of the Forest.

## 1.2. Study Area

Mago National Park is one of the parks in Ethiopia located in SNNPR of Ethiopia. It is about 782 km south of Addis Ababa and north of a large 90° bend in the Omo River. The Mago national park was established in 1979. The Mago Park covers area about 1869.95 km2. Geographically, the park lies between latitude 05°20'-05°50'N and longitude 36°00'-36°30'E. The elevation ranges from 400m.a.s.l on the plains in south, to 1,776 m on top of Mt Mago. The interior section of the park mainly consists of flat plains. However, periphery and boundaries, except to the south, are formed by the Mago and Mursi Mountains, associated ridges and chains of hills

*Editor's note: the figure was removed due to copyright issues.*

Figure 1.Map of the study area

## 2. Key findings and Discussion

### 2.1 An Overview of the Reviewed Articles related to Allometric Equation

Allometric equations have been developed for different types of vegetation in Ethiopia. This review tries to concentrate mostly on recent articles (Figure 2.1). Most developed equations were undertaken since 2016.

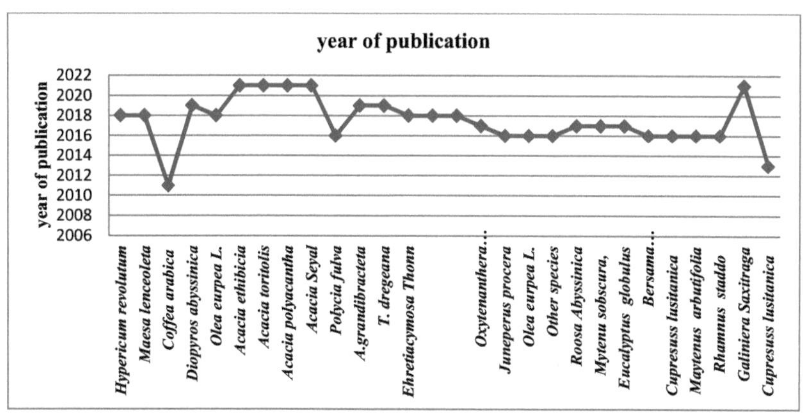


Fig 2.1 Year of publication for articles reviewed

IPCC Recommend that for Allometric equation development a minimum of thirty plant species should be selected (IPCC, 2006). 57% of the reviewed articles selected 12 plant species which is below half of the recommended. Only 29% reviewed articles selected appropriate sample (Fig 2.2). The one which selected 45 plant species was aimed to develop Allometric equation for mult- species or the Allometric equation developed was site specific but not species specific. From the reviewed articles 14% on Acacia Species and this make largest share from selected plant Species.

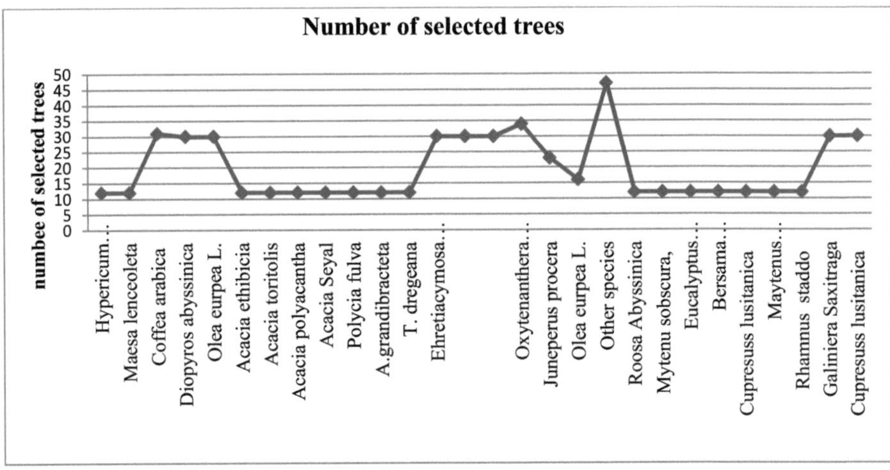


Fig 2.2 Number of selected trees for Allometric equation development for each plant species

Destructive method is the most accurate method to develop regression equations from destructively sampled trees that are in the size range of interest (Abola et al., 2005). According to IPCC, (2006) the increasing order of accuracy to develop regression equation are Destructive, semi-destructive and using default value. Despite this fact most of the article reviewed (75%) use semi- destructive method for Allometric equation development. Only the rest percentage use destructive method.

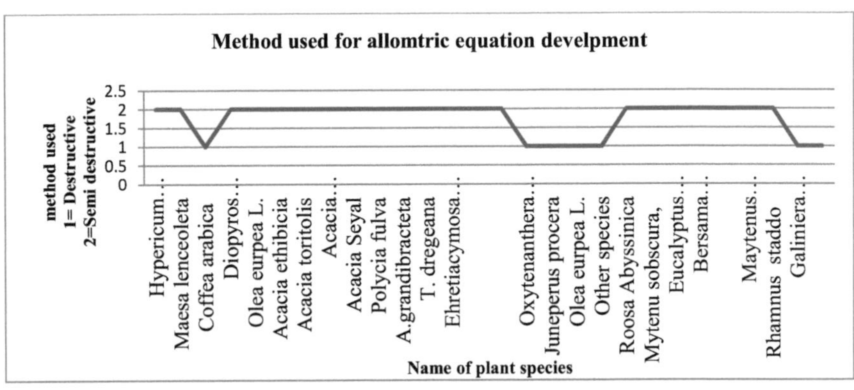

Fig. 2.3 Method used for Allometric equation development by reviewed articles

## 2.2 Major Findings (What is known) on Allometric Equation

Several study have been conducted on the development of allometric equation in Ethiopia. Most of them were conducted by using semi destructive method. The following are some of the conducted research in Ethiopia with the relationship between biomass and dendrometric variables. Mecheal Hordofa et al., (2018) conducted a research in Chuqala natural forest and develop allometric models for biomass estimation with independent observable predictor, such as Diameter at Breast Height (DBH) of a tree, height of a tree, and volume of tree components. According to their result, among the independent variables that are DBH, H, and V the most appropriate independent variable to fit the models for the two selected tree were the DBH. Birhanu Kebede &Teshome Soromessa (2018) Pointed out that, the AGB was strongly correlated with DBH and was not significantly correlated with wood density and height individually in *O. europaea* L. subsp. cuspidata allometric equation development. In combination, AGB was strongly correlated with DBH and height; DBH and wood density; and the combination of DBH, height, and wood density. They also indicate that the Species-specific equations are better carbon assessment than general equations. Another study on Allometric equation development was the study undertaken by Mulugeta Mokria et al., (2018). They harvested and weighed 84 trees from eleven dominant species from six grazing enclosures and adjacent communal grazing land. They observed that AGB correlates significantly with diameter at stump height DSH, and tree height H.

Abiy Mulat and Teshome Soromessa, (2016) also indicate that, DBH alone is a good predictor of biomass especially, in terms of multiple tradeoffs between accuracy, cost and practicability of the measurement. And also they recommended that the use of model where tree biomass is determined from DBH only, which had a practical advantage because most of the inventories include DBH measurements. Moreover, DBH is easy to measure accurately in the field.

Another study which was done in the Banja district of Awi zone estimate AGB against combination dendrometric variables such as DBH, H and D in the species specific allometric model showed the best model performance. The developed models related AGB against predictors in the log–log forms and these model relationships between AGB and predictors was the best in terms of goodness of fits. Their result revealed that the inclusion of diameter, height and wood density in the development of species-specific allometric equations predicted aboveground biomass with small bias than using a single or two predictors (Getaneh et al., 2020)

The allometric equations are used to predict tree and stand biomass, based on easily measured tree variables such as DBH and height. Tesfaye et al., (2017) pointed out that DBH was only explanatory variable provides a satisfactory estimation of biomass since the total variation explained by the relationship is high and the associated bias was small. Their revealed also as DBH is a strong indicator of above ground biomass. So, DBH alone is a good predictor of biomass especially, in terms of multiple tradeoffs between accuracy, cost and practicability of the measurement. The study by Yehualashet et al., (2016) Egdu Forest indicates that diameter at breast height, wood specific gravity, and tree height were important estimator to consider for the estimation of aboveground biomass at tree scale. Comparison of these results with those obtained using generalized allometric model revealed differences with biomass estimation. The application of the newly proposed allometric models from this study shall be based on the availability of forest inventory data for incorporation of the various estimator variables.

Mesele Negash et al., (2013) *Coffea arabica* L.grown in the Rift Valley escarpment of Ethiopia. The harvests 31 plants and develop power equations using stem diameter measured at either 40 cm (d40) or at breast height (d, 1.3 m) with and without stem height (h) were evaluated. The square power equation, $Y = b_1d^2 40$, was found to be the best (highest ranked using goodness-of-fit statistics) for predicting total and component biomass. The study conducted in the Omo-Gibe woodland of south-western Ethiopia to develop an allometric equation to estimate the Above-ground Biomass (AGB) of the four Acacia species (*Acacia polyacantha, Acacia seyal, Acacia etbaica* and *Acacia tortilis*). Fifty- four (54) acacia trees were sampled According to their result the model containing DBH alone was more accurate to estimate AGB compared to the use of multiple predictor variables (Abreham Berta et al., 2021). Fekadu Gurmessa et al., (2017) conduct a research in Dicho forest to develop Allometric Equations to Estimate the Biomass of *Oxytenanthera Abyssinica* (A. Rich.) They develop linear allometric equations to estimate TAGB and TBGB of individual bamboo trees. In order to develop the regression equation they used destructive method. Accordingly, the linear model in the form of Y = a + b*X (where: Y is biomass, a and b are constant parameters, and X is independent variable such as DBH, Height, Basal diameter and/or basal area or a combination of these variables) They were tried many possibilities using DBH, height, basal diameter (D10), basal area (BA), wood density ($\varrho$) and their combinations and logarithmic transformations as predictor variable. They also found that DBH was the best predictor variable for the total aboveground biomass (TAGB). Another study conducted by Buruh Abebe et al., (2016) in Desa'a dry Afromontane forest was by using destructive methods. They tried to develop Allometric equation for two plant species namely *Juneperus procera* and *Olea eurpea* L and another equation for the rest plant species. They identify relationships between the response variable (AGB) and the potential predictor variables (DBH, DSH, H, CA, $\rho$). Among all the tested AGB models for two dominant species and for other species under Model System II, only those with DBH and DSH as sole predictor variables had significant parameter estimates. This article was the only article that I reviewed that use DSH for large trees. In 2021 Tura Bareke, and Admassu Addi conducted study in Gesha-Sayilem forest, on allometric equation for aboveground biomass estimation of *Galiniera saxifraga* (Hochst.) Their results indicated that DBH and crown area were found to be the best fit variables for *G. saxifrage*. This was the only study that I came across thatuse crown area for developing Allometric equation.

Table 2.1 Summary of literature Reviewed which include Species name, Equation and Area in which the study undertaken

| s/n | Species | Equation | Area of study |
|---|---|---|---|
| 1 | *Hypericum revolutum* | $AGB = -681.015 + 4{,}494.094\,(DBH)$ | Chuqala Natural Forest |
| 2 | *Maesa lanceoleta* | $AGB = -936.96 + 5{,}268.92\,(DBH)$ | Chuqala Natural Forest |
| 3 | *Coffeaarabica* L. | $Y = 0.147 D_{40}^{2}$ | Rift Valley escarpment of Ethiopia |
| 4 | *Olea europaea* L. | $AGB_{est} = 0.623 \times (D)^{1.352} \times (H)^{0.703}$ and $AGB_{est} = 0.866 \times (DBH)^{1.432} \times (H)^{0.608} \times (\rho)^{1.067}$ | ManaAngetu Forest |
| 5 | *Albizia grandibracteata* | $TAGB = 0.3274 \times (\rho D2H)^{0.759}$ | Yayu Coffee-Forest Biosphere |
| 5 | *Trichilia dregeana* | $TAGB = 0.0832 \times (\rho D2H)0.899$ | |
| 6 | multi-species allometric equation | $Y = 0.2655 * (D_{30})^{1.7737}$ | 'Gomit watershed', located in South Gondar zone, |
| 7 | *Polysciasfulva*(indigenous tree). | $AGB = -301.621 + 14.056\,(DBH)$ and $\ln(AGB) = \ln(-2.439) + 2.129\,\ln(DBH)$ | In Agroforestry Gedeo Zone of Ethiopia |

| | | | |
|---|---|---|---|
| | | Total biomass=-361.945 +16.868 (DBH) | |
| 8 | *Diospyrosabyssinica* (Hiern) F. White tree Species | $AGB_{est} = 0{:}457 \times D^2 H^{0.721} \times (\rho)^{0.446}$ | Yayu Coffee-Forest Biosphere |
| 9 | *Acacia seyal* | $Ln\,(AGB) = 2.20636 + 0.53167 \ln(DBH^2)$ | Selected Indigenous Acacia Species in Omo-Gibe Woodland Ecosystem |
| 10 | *Acacia polyacantha* | $\ln(AGB) = 2.95854 + 0.21750\ln((DBH)^2 H\rho)$ | |
| 11 | *Acacia ethibicia* | $AGB = 29.01898 \times ((DBH)^2 H\rho)^{0.21518}$ | |
| 12 | *Acacia toritolis* | $AGB = 3.82427 \times ((DBH)^2 H\rho)^{0.26748}$ | |
| 13 | *Eucalyptus globules* | $AGB = ((0.08283 \times (DBH^{1.873})) \times (H^{0.8242}) + 10^{(-3)}$ | Gambo forest |
| 14 | *Ehretiacymosa* Thonn | $AGB = \exp[-1.137 + 2.129 \ln(DBH) - 0.035 \ln(H) + 0.460 \ln(WD)]$ | Banja district Awi zone |
| 15 | *Lepidotrichilia volkensii (Giirke) Leroy* | $AGB = \exp[-0.040 + 1.758 \ln(DBH) - 0.094 \ln(H) + 0.519 \ln(WD)].$ | |
| 16 | *Rothmannia urcelliformis*(Hiern) | $AGB = \exp[-2.309 + 2.289 \ln(DBH) + 0.019 (H) - 0.937 \ln(WD)].$ | Banja district Awi zone |
| 17 | *Tecleanobilis*Del. | $AGB = \exp[-1.290 + 1.667 \ln(DBH) + 0.397 \ln(H) - 0.639 \ln(WD)].$ | |
| 18 | *Allophyllusabyssinicus* | $W_{above} = \beta \times (d \times h)$ | |
| 19 | *Oleaeuropaeassp.cuspidata* | $W_{above} = \Sigma\ \beta \times (d^2 \times h) + (\beta \times d2) + (\lambda \times h) + \beta \times (d2 \times h)$ | Chilimo-Gaji dry afro-montane forest |
| 20 | *Oliniarochetiana* | $W_{above} = \Sigma\ \beta \times (d \times h) + (\beta \times d^2) + \lambda \times (d^2 \times h)$ | |
| 21 | *Rhus glutinosa* | $W_{above} = \Sigma\ \beta \times (d \times h) + (\beta \times d^2) + (\lambda \times h)$ | |
| 22 | *Scolopi atheifolia.* | $W_{above} = \Sigma\ \beta \times (d^2 \times h) + \beta \times (d \times h)$ | |
| 23 | *Oxytenanthera abyssinica*(A. Rich.)(Ethiopian Lowland Bamboo) | $TAGB = -1.073 + 0.067 \times BA + \quad -0.064 \times D\mathbf{10} + 0.206 \times H + 0.653 \times DBH$ | Dicho Forest |
| 24 | *Crotonmacrostachys* Hochst. Del. | $y = 22.601 DBH - 242.74$ | Munessa-Shashemene Forest Degaga District |
| 25 | *Cupressus lusitanica Miller, Cupressaceae* | $y = 27.293 DBH - 380.14$ | |
| 26 | *Eucalyptus globulus Labill., Myrtaceae* | $y = 29.517 DBH - 294.8$ | |
| 27 | *Eucalyptus globules* | $AGB\ best = 0.295 \times (DBH)^{2.088}$ | Egdu Forest, in Welmera District |
| 28 | *Maytenus obscura* | $AGB\ best = 0.255 \times (DBH)^{2.103}$ | |
| 29 | *Rosa abyssinica* | $AGB\ best = 0.371 \times (DBH)^{2.0251}$ | |
| 30 | *Juniperus procera* | $\ln(AGB) = \ln(2.48) + 2.321\ln(DBH)$ | Wof-washa forest in Ethiopia |
| 31 | *Bersama abyssinica* | $AGB = 9.996 + 0.518(DBH) - 0.044(H) - 17.37(\rho)$ | Egdu Forest, in Welmera District |
| 32 | *Cupressus lusitanica,* | $AGB = -193.359 + 25.869(DBH) - 15.727(H) + 90.952(\rho)$ | |
| 33 | *Maytenus arbutifolia,* | $AGB = 5.538 + 1.9545(DBH) + 0.316(H) + 8.01(\rho)$ | |

| 34 | *Rhamnus staddo* | AGB=-2.25+3.220(DBH)-1.356(H) | |
| 35 | *Cupressus lusitanica Miller, Cupressaceae* | $Y = -2.8157D^2H + e$ | Wondo Genet Plantation |
| 36 | multi-species allometric equation | $AGB = 0.298 + DBH^{2.034}$ | Dry Afromontane Forests of Northern Ethiopia |

## 2.3 Major Findings (What is known) on Carbon Sequestration on Agroforestry

The capacity of Agroforestry systems to store Carbon varies among different agro-ecological landscapes and the types of Agroforestry system (Montagnini, F. and Nair, P.K.R., 2004). As a result, the storage potential for semiarid, sub-humid, humid, and temperate regions is estimated at 9, 21, 50, and 63 t C ha$^{-1}$, respectively (Montagnini, F. and Nair, P.K.R., 2004). Extensive reviews by (Luedeling, E. and Neufeldt, H., 2012) for West African Sahel countries (extending from arid Sahara Desert to humid region Guinea) reported biomass Carbon stocks ranging from 22.2 to 70.8 t C ha$^{-1}$. As reports showed, globally, the total biomass Carbon stock for AF systems ranges between 12 and 228 t C ha$^{-1}$(Albrecht, A. and Kandji, S.T., 2003). Another study by (Nair, P.K. and Nair, V.D., 2003) reported that Agroforestry practices stored Carbon ranging from 0.29 to 15.21 t C ha$^{-1}$ yr$^{-1}$ in their aboveground biomass and can have from 30 to 300 t C ha$^{-1}$ in their soil down to one-meter depth. Soil Carbon stock for the 0–60 cm soil layer differs among different land uses and regions. For instance, the Carbon stock in the above mentioned soil layer is 121–123 t ha$^{-1}$ for tropical forests and 110–117 t ha$^{-1}$ for tropical savanna (Lal, R., 2004). ). As reported by Takimoto et al. (2008) in the study of an agroforestry systems (traditional parkland systems with Faidherbia albida and Vitellaria paradoxa as the dominant tree species, live fence, fodder banks and an abandoned previously cultivated land) in the West African Sahel, biomass Carbon stock ranged from 0.7 to 54.0 Mg C ha$^{-1}$ and total Carbon stock (biomass Carbon + soil Carbon, into 1m depth) from 28.7 to 87.3 Mg C ha-1, in which a major portion of the total amount of C in the system is stored in the soil. The studies conducted in Ethiopia also showed that the Agroforestry systems have a great potential in sequestering a significant quantity of Carbon. For instance: Reta Eshetu et al., (2015) Undertaken a research on Parkland Agroforestry Practice in Minjar Shenkora. Their result revealed that, the average, carbon stock of parklands practice in Minjar shenkora was 59.65 Mg C ha$^{-1}$. Another study  by Mihert Semere in (2019) on Biomass and Soil Carbon Stocks Assessment of Agroforestry Systems  and Adjacent Cultivated Land, in Cheha Wereda, reported that the total ecosystem carbon stocks in home garden and woodlot AFS were 100.4 and 72.9Mg C ha-1 respectively.

This study also showed that the highest SOC stock was recorded in home garden agroforestry system. The study undertaken on Assessing carbon pools of three indigenous Agroforestry systems in the Southeastern Rift-Valley Landscapes, Ethiopia, indicated that the mean AGB ranged from 81.1 to 255.9 t ha$^{-1}$ and for BGB from 26.9 to 72.2 t ha$^{-1}$. The highest C stock was found in Coffee–Fruit tree–Enset based (233.3 $\pm$ 81.0 t ha$^{-1}$), and the lowest was in Coffee–Enset based AF system (190.1 $\pm$ 29.8 t ha$^{-1}$). The carbon stock in Enset based AF system was (197.8 $\pm$ 58.7 t ha$^{-1}$) (Tesfay, H.M. et al., 2022). Negash, M. and Starr, M., 2015 undertaken the study on biomass and soil carbon stocks of indigenous agroforestry systems on the south-eastern Rift Valley escarpment. Their result indicated that the smallholding total biomass C stocks averaged 67 Mg ha$^{-1}$ and SOC stocks (0–60 cm) were 109–253 Mg ha−1 (52–91% of total C stocks). As reported by Ashenafi et al,. (2021) the four smallholdings AF practices (parkland, home garden, boundary planting and woodlot) Carbon stocks ranged from 77 to 135 Mg ha$^{-1}$.

## 2.4 Major Findings (What is known) on application of general Allometric equation

Allometric equations are important for their application to local and national forest carbon assessments, as well as for global carbon balance assessments (Basuki TM., et al., 2009). Primarily, the current issue of global carbon cycles is the prominent factor for the formulation of biomass regression models (Henry M., et al., 2011, Jara MC., et al., 2014, Wang C., 2006). As a result, generalized pantropical allometric equations were developed by many researchers (Henry M., et al., 2011, Chave J., et al., 2005, Brown S. 1997). The development of a generalized allometric equation was approached by measuring multiple tree species and it was intended to be applied to a broad range of tropical forests (Chave J., et al., 2005). However, a great error is generated related to adopting generic pantropical allometric equations for many forests (Alvarez E. et al., 2012, Ngomanda A. et al., 2014). Biomass error that can be generated at individual tree level is also regularly propagated bias at forest stand and country level during the assessment of biomass and carbon stock change the appropriate allometric equation is not used.

Environmental variations among different forests are the ultimate factors for the variation of their biomass. Climatic regimes are the prominent factors that affect the growth of woody plants and biomass accumulation of different forest stands (Xu X. et al., 2018, Pfeifer M. et al., 2018). Also, environmental variability in the context of physiographic and edaphic conditions plays a significant role in the variation of species composition and biomass difference among different forest sites . (Alves LF. et al., 2010, Laumonier Y. et al., 2010). Within-stand variation of biomass for different tree species is related to tree architecture, growth strategies and its dynamic interplay with the biophysical environments (Ketterings QM. et al., 2001, Muller-Landau HC., 2004, Clark DB and Clark DA. 2000.]. The difference of TAGB across a forest landscape is mostly related to the variation in slope, elevation, and aspect (Alves LF. et al., 2010, SalinaS-Melgoza MA. et al.,  2018). Generally, tropical forests are known for their high diversity of woody plants. The application of multispecies pan-tropical equations to individual tree species generates uncertainty of TAGB (Basuki TM., et al., 2009, Djomo AN and Chimi CD. 2017). Therefore, formulating a species-and site-specific biomass regression model was found the best approach to accurately quantify biomass and carbon storage of forests (Goussanou C. et al., 2016).

## 2.5. What is not known/ New

Species specific models provide less bias than general models and site specific because, local climatic, soil properties, altitude, and land-use history are affected by tree growth characteristics (Yuen J.Q. et al., 2016).  According to Mokria M. et al., (2018), the lack of a species -specific allometric model for estimation of AGB is the key reason for persistent inaccuracy and low uncertainty in biomass estimation in sub-Saharan Africa. Solomon et al., (2012) stated that the precise estimation of biomass and carbon stock in a forest can be achieved using species specific and site-specific allometric equations for the species and forest types. According to Henry M. et al., (2019), reviews of biomass models in sub-Saharan Africa, 63 models have been established in Ethiopia, which focused only on six allometric equations, of which 70% of the models were developed for eucalyptus species (Tetemke et al., 2019). However, following the review of Henry M. et al., (2011), there have been attempts to develop local species-specific and site specific allometric equations for estimating AGB of trees in different parts of Ethiopia in recent studies such as Tetemke et al., (2019), Feyisa, K., et al., (2018), Kebede and Soromessa (2018), Daba and Soromessa (2019) and  Daba and Soromessa (2019) but still, they are not sufficient in respective of vegetation types and agro-ecology  of Ethiopia. In addition to this fact most the species specific Allometric equations developed depend on semi-destructive which was less accurate than destructive method. The development of a new allometric equation requires that at least 30 trees covering the full range of diameter classes. If the regression based on the 30 trees does not result in a statistically significant relationship (high r-squared value), then additional trees will need to be harvested (Walker SM et al., 2015). According to the reviewed literature, most of the study took sampled trees which were less than the minimum required for the investigation.

Many of adopted generalized equations generate great uncertainty of biomass. The study of (Alvarez E. et al., 2012) reports that higher bias was observed related to the Chave's model II largely overestimating by approximately 300% to 400% for two tropical forest sites. This confirms the significance of formulating species-and site-specific allometric equations for tropical forests. Ethiopia is known for its diverse vegetation ecosystems and associated high diversity of woody plants. However, the assessment of biomass and carbon stock of forests has been practiced by adopting the generic pan-tropical allometric equations that cause great uncertainty. This uncertainty can be recognized by comparing measured value with species specific and general allometric equation.

The different studies have been conducted by different researchers in Ethiopia regarding carbon sequestration potential of Agroforestry such as, Reta Eshetu et al,. (2015), Ashenafi et al,. (2021), Tesfay, H.M. et al., (2022), Seta, T.; Demissew, S., (2014), Negash, M. and Starr, M., (2015), Mihert Semere, (2019), Birhane, E. et al., (2020), and Siyum, G.E. and Tassew, T. (2019) but this study cover only small portion if our country so, further study is important.

In the case of my review area, no allometric equation has been developed for selected tree species. Regarding Agroforestry there was a study which was conducted by (Wele, D. Ambachewet al., 2009) on carbon stocks in traditional Agroforestry systems adjacent to Mago Forest.


## 3. CONCLUSION.

United Nations Framework Convention on Climate Change recently agreed to study and consider new initiatives, in favor of forest-rich developing countries. It provides financial or economic incentives to help developing countries voluntarily to reduce national deforestation rates and associated carbon emissions below a baseline (REDD). Countries that demonstrate emissions reductions may be able to sell those carbon credits to the international carbon market or elsewhere. Under the UNFCCC, countries have to regularly report the state of their forest resources. Forest ecosystem and Agroforestry system are the major constituent of the carbon reserves and it plays an important role in regulating global climate change through process of carbon sequestration. Measuring biomass in local, regional and global scales is critical for estimating global carbon storage and assessing ecosystem response to climate change. Formulating allometric equations for all woody plants in Ethiopia is quite desirable for accurately quantifying the biomass and carbon stock of forests to achieve accurate national and international reporting of carbon dioxide emission inventories. Depending on this in Ethiopia several study undertaken regarding allometric equation for biomass estimation, but it is not as forest resource of the country. Most of the study undertaken was depend on semi-destructive method which is not accurate as destructive measurement. The number of plant selected for biomass estimation by allometric equation was much more less than the standard put by IPCC. Generally, in response to global climate change mitigation, the monitoring and assessment of carbon dioxide from forests and Agroforestry is essential.

## 4. REFERENCE


Abiy Mulat and Teshome Soromessa, 2016. Species Specific Allometric Model for Biomass Estimation of Polyscias fulva Harms in Tumata Chirecha Agroforestry Gedeo Zone of Ethiopia: Implication for Sustainable Management and Climatic Change Mitigation *Journal of Energy Technologies and Policy,* ISSN 2224-3232 (Paper) ISSN 2225-0573 (Online)

Abreham Berta Aneseyee, Teshome Soromessa, Eyasu Elias, and Gudina Legese Feyisa, 2012. Allometric Equations For Ethiopian Acacia Species Center for Environmental Science, College of Computational and Natural Science, Addis Ababa University, P. O. Box No: 1176, Addis Ababa, Ethiopia https://orcid.org/0000-0002-3076-3768

Addi A, Demissew S, Soromessa T, Asfaw Z., 2019. Carbon stock of the moist Afromontane forest in Gesha and Sayilem Districts in Kaffa Zone: An implication for climate change mitigation. J Ecosyst Ecograph 9 (1): 1-8.

Albrecht, A.; Kandji, S.T. Carbon sequestration in tropical agroforestry systems. Agric. Ecosyst. Environ. **2003**, 99, 15–27. [CrossRef]

Ali A, Xu MS, Zhao YT, Zhang QQ, Zhou LL, Yang XD, Yan ER., 2015. Allometric biomass equations for shrub and small tree species in subtropical China. Silva Fenn 49 (4): 1-10. DOI: 10.14214/sf.1275.

Alvarez E, Duque A, Saldarriaga J, Cabrera K, de las Salas G, del Valle I,  2012. Tree above-ground biomass allometries for carbon stocks estimation in the natural forests of Colombia. For Ecol Manag.  267:297–308.

Alves LF, Vieira SA, Scaranello MA, Camargo PB, Santos FAM, Joly CA, et al. Forest structure and live aboveground biomass variation along an elevational gradient of tropical Atlantic moist forest (Brazil). For Ecol Manag. 2010;260(5):679–91.

Ancelm MW, Edward ME, Luoga E, Ernest MR, Zahabu E, Santos SD, Sola G, Crete P, Henry M, Kashindye A., 2016. Allometric models for estimating tree volume and aboveground biomass in Lowland Forests of Tanzania. Int J For Res 2016 (4): 1-13. DOI: 10.1155/2016/8076271.

Ashenafi Manaye, Berihu Tesfamariam, Musse Tesfaye, Adefires Worku and Yirga Gufi. 2021 Tree diversity and carbon stocks in agroforestry systems in northern Ethiopia

Basuki TM, van Laake PE, Skidmore AK, Hussin YA. 2009. Allometric equations for estimating the above-ground biomass in tropical lowland Dipterocarp forests. For Ecol Manag.;257(8):1684–94.

Brassard BW, Chen HYH, Bergeron Y., 2009. Influence of environmental variability on root dynamics in northern forests. Crit Rev Plant Sci 28: 179-197.

Birhanu Kebede and Teshome Soromessa,  2018. Allometric equations for aboveground biomass estimation of Olea europaea L. subsp. cuspidata in Mana Angetu Forest. Ecosyst Health Sustain. 2018;4 (1):1–12.

Birhane, E.; Ahmed, S.; Hailemariam, M.; Negash, M.; Rannestad, M.M.; Norgrove, L.2020. Carbon stock and woody species diversity in homegarden agroforestry along an elevation gradient in southern Ethiopia. Agrofor. Syst, 94, 1099–1110. [CrossRef]

Brown S. 1997. Estimating biomass and biomass change of tropical forests: a primer. Rome: FAO Forestry Pape. p. 134.

Brown, S., 2002. Measuring carbon in forests: current status and future challenges. Environ. Pollut. 116, 363–372.

Buruh Abebe Tetemke, Emiru Birhane, Meley Mekonen Rannestad and Tron Eid, 2016. Allometric Models for Predicting Aboveground Biomass of Trees in the Dry Afromontane Forests of Northern Ethiopia

Chave, J., Condit, R., Aguilar, S., Hernandez, A., Lao, S., Perez, R., 2004. Error propagation and scaling for tropical forest biomass estimate. Philosophical Transactions of the Royal Society of London Series B 359: 409–420.

Chave, J., Andalo, C., Brown, S., Cairns, A., Chambers, Q., Eamus, D., Fölster, H., Fromard, F., Higuchi, N., Kira, T., Lescure, P., Nelson, W., Ogawa, H., Puig, H., Riéra, B., and Yamakura, T., 2005. Tree allometry and improved estimation of carbon stocks and balance in tropical forests. Oecologia.; 145:87–99.

Chave, J., M. Rejou-Mechain, A. Burquez, E. Chidumayo, M. S. Colgan, W. B. Delitti, A. Duque, et al., 2014. "Improved Allometric Models to Estimate the Aboveground Biomass of Tropical Trees." Global Change Biology 20 (10): 3177–3190. doi:10.1111/ gcb.12629.

Clark DB, Clark DA. 2000. Landscape-scale variation in forest structure and biomass in a tropical rain forest. For Ecol Manag. 137:185–98.

Damena Edae Daba & Teshome Soromessa, 2019. Allometric equations for aboveground biomass estimation of Diospyros abyssinica (Hiern) F. White tree species, Ecosystem. Health and Sustainability, 5:1, 86-97, DOI: 10.1080/20964129.2019.1591169.

Damena Edae Daba and Teshome Soromessa, 2019. The accuracy of species-specific allometric equations for estimating aboveground biomass in tropical moist montane forests: case study of Albizia grandibracteata and Trichilia dregeana, Carbon Balance Manage 14:18 https://doi.org/10.1186/s13021-019-0134-8

Djomo AN, Chimi CD. 2017.Tree allometric equations for estimation of above, below and total biomass in a tropical moist forest: case study with application to remote sensing. For Ecol Manag. ;391:184–93.

Dixon, R. K. 1995. Agroforestry systems: sources of sinks of greenhouse gases? Agroforestry Systems 31:99-116.

Fayolle, A., Doucet, J.L., Gillet, J.F., Bourland, N., Lejeune, P., 2013. Tree allometry in Central Africa: testing the validity of pantropical multi-species allometric equations for estimating biomass and carbon stocks. Forest Ecology and Management 305:29–37.

FAO (Food and Agriculture Organization), 2012. Manual for building tree volume and biomass allometric equations: from field measurement to prediction00153 Rome, Italiepp 51-78 august.

Feyisa, K., 2018. Allometric equations for predicting above-ground biomass of selected woody species to estimate carbon in East African rangelands. Agroforestry Systems, 92(3): p. 599-621.

Fekadu Gurmessa, Teshome Gemechu, Teshome Soromessa, and Ensermu Kelbessa, 2017. Allometric Equations to Estimate the Biomass of Oxytenanthera Abyssinica (A. Rich.) Munro. (Ethiopian Lowland Bamboo) in Dicho Forest, Oromia Region, Western Ethiopia International Journal of Research Studies in Biosciences (IJRSB) Volume 4, Issue 12, December 2016, PP 34-48

Gebremeskel, D.; Birhane, E.; Rannestad, M.M.; Gebre, S.; Tesfay, G. 2001. Biomass and soil carbon stocks of Rhamnus prinoides based agroforestry practice with varied density in the drylands of Northern Ethiopia. Agrofor. Syst. 95, 1275–1293.

Getaneh Gebeyehu, Teshome Soromessa, Tesfaye Bekele, and Demel Teketay, 2016. Allometric Equations for Aboveground Biomass Estimations of Four Dry Afromontane Tree Species

Goussanou C, Guendehou S, Assogbadjo A, Kaire M, Sinsin B, Cuni- Sanchez A.2016. Specific and generic stem biomass and volume models of tree species in a West African tropical semi-deciduous forest. Silva Fennica. 50(2):1474.

Haile, S.G.; Nair, P.K.R.; Nair, V.D. 2008. Carbon Storage of Different Soil-Size Fractions in Florida Silvopastoral Systems. J. Environ. Qual. 37, 1789–1797.

Han FX, Plodinec MJ, Su Y, Monts DL, Li Z (2007) Terrestrial carbon pools in southeast and South-Central United States. Clim Chang 84:191–202.

Henry, M., Picard, N., Trotta, C., Manlay, R.J., Valentini, R., Bernoux, M., Saint-André, L., 2011. Estimating Tree Biomass of Sub-Saharan African Forests: Review of available allometric equations .Silva Fennica 45: 477–569.

Intergovernmental Panel on Climate Change (IPCC). 2006. IPCC Guide lines for National Greenhouse Gas inventories. Prepared by the National Greenhouse Gas Inventories program, Eggleston HS, Buenidias L, Miwak K, Ngara T, Tanabek K, eds. Published: IGES, Japan.

Jara MC, Henry M, Réjou-Méchain M, Wayson C, Zapata-Cuartas M, Piotto D,  2014. Guidelines for documenting and reporting tree allometric equations. Ann For Sci. 72(6):763–8.

Kale M, Singh S, Roy PS, Deosthali V, Ghole VS (2004) Biomass equations of dominant species of dry deciduous forest in Shivpuri district, Madhya Pradesh. Curr Sci 87(5):683–687.

Ketterings, Q.M.,Coe, R., Noordwijk, M., Ambagau, Y., Palm, C.A., 2001. Reducing uncertainty in the use of allometric biomass equations for predicting above-ground tree biomass in mixed secondary forests. For. Ecol. Manage 146:199– 209.

Lal, R. 2004. Soil carbon sequestration to mitigate climate change. Geoderma. 123, 1–22. [CrossRef]

Laumonier Y, Edin A, Kanninen M, Munandar AW. 2010. Landscape-scale variation in the structure and biomass of the hill dipterocarp forest of Sumatra: Implications for carbon stock assessments. For Ecol Manag. 259(3):505–13.

Luedeling, E.; Neufeldt, H. Carbon sequestration potential of parkland agroforestry in the Sahel. Clim. Chang. **2012**, 115, 443–461. [CrossRef]

Mecheal Hordofa Balcha, Teshome Soromessal and Dejene Kebede, 2018. Allometric Equation for Biomass Determination in Chuqala Natural Forest, Ethiopia: Implication for Climate Change Mitigation, Journal of Forest and Environmental Science http://jofs.or.kr Vol. 34, No. 2, pp. 108-118.

Mihert Semere, 2019 Biomass and Soil Carbon Stocks Assessment of Agroforestry Systems and Adjacent Cultivated Land, in Cheha Wereda, Gurage Zone

MLOH (Mauna Loa Observatory in Hawaii), 2015. http://www.esrl.noaa.gov/gmd/obop/mlo

Mokria M, Mekuria W, Gebrekirstos A, Aynekulu E, Belay B, Gashaw T, Bräuning A., 2018. Mixed-species allometric equations and estimation of aboveground biomass and carbon stocks in restoring degraded landscape in northern Ethiopia. Environ Res Lett 13: 024022.

Montagnini, F.; Nair, P.K.R. 2004. Carbon sequestration: An underexploited environmental benefit of agroforestry systems. Agrofor. Syst. , 61–62, 281–295.

Muller-Landau HC. 2004. Interspecific and inter-site variation in wood specific gravity of tropical trees. Biotropica. 36(1):20–322.

Návar J, Nájera J, Jurado E., 2002. Biomass estimation equations in the Tamaulipan thornscrub of north-eastern Mexico. J Arid Environ 52: 167-179.

Nair, P.K. and Nair, V.D. 2003. Carbon storage in North American agroforestry systems. In The Potential of U.S. Forest Soils to Sequester Carbon and Mitigate the Greenhouse Effect; Kimble, J., Heath, L.S., Birdsey, R.A., Lal, R., Eds.; CRC Press: Boca Raton, FL, USA; pp. 333–346.

Nair, P.R.; Nair, V.D.; Kumar, B.M.; Haile, S.G. 2009. Soil carbon sequestration in tropical agroforestry systems: A feasibility appraisal. Environ. Sci. Policy. 12, 1099–1111.

NEFA (2002) North East State Forest Association. Carbon sequestration and its impacts on forest management in the northeast. Available at http://www.nefainfo.org/publications/carbonsequestration.pdf.

Negash M., Mike Starr M., Markku Kanninen M., and Berhe L., 2013. Allometric equations for estimating aboveground biomass of Coffea  arabica L. grown in the Rift Valley escarpment of Ethiopia. Agroforestry systems, 87(4): p. 953-966.

Negash, M.; Starr, M. 2015. Biomass and soil carbon stocks of indigenous agroforestry systems on the south-eastern Rift Valley escarpment, Ethiopia. Plant Soil, 393, 95–107. [CrossRef]

Ngomanda A, Obiang Obiang NL, Lebamba J, Moundounga Mavouroulou Q, Gomat H, Mankou GS, 2014. Site-specific versus pantropical allometric equations: which option to estimate the biomass of a moist central African forest? For Ecol Manag. 312:1–9. Ngomanda A., et al., 2014

Pfeifer M, Gonsamo A, Woodgate W, Cayuela L, Marshall AR, Ledo A,  2018. Tropical forest canopies and their relationships with climate and disturbance: results from a global dataset of consistent field-based measurements. For Ecosyst. 2018;5(1):7. Pfeifer M. et al., 2018

Picard, N., Saint André, L., Henry, M., 2012. Manual for building tree allometric equations: from the field to the estimation, Food and Agriculture Organization of the United Nations, Centre de Coopération Internationale en Recherche Agronomique.

Reta Eshetu,  Seid Muhie, and solomon Mulu, 2015. Assessment of Carbon Stock Potential of Parkland Agroforestry Practice: The Case of Minjar Shenkora; North Shewa, Ethiopia

SalinaS-Melgoza MA, Skutsch M, Lovett JC. 2018. Predicting aboveground forest biomass with topographic variables in human-impacted tropical dry forest landscapes. Ecosphere. 9(1):1–20.

Sharrow, S.; Ismail, S. 2004. Carbon and nitrogen storage in agroforests, tree plantations, and pastures in western Oregon, USA. Agrofor. Syst. 60, 123–130.

Seta, T.; Demissew, S. 2014. Diversity and standing carbon stocks of native agroforestry trees in Wenago district, Ethiopia. J. Emerg. Trends Eng. Appl. Sci. (JETEAS), 5, 125–132

Siyum, G.E.; Tassew, T.2019.  The Use of Homegarden Agroforestry Systems for Climate Change Mitigation in Lowlands of Southern Tigray, Northern Ethiopia. Asian Soil Res. J, 2, 1–13. [CrossRef]

Slik, F., Bernard, S., Breman, C., Beek, M., Salim, A., Sheil, D., 2008. Wood density as aconservation tool: quantification of disturbance and identification of conservation priority areas in tropical forests. Conservation Biology 22(5):1299-1308

Solomon N, Birhane E, Tadesse T, Treydte AC, Meles K., 2017. Carbon stocks and sequestration potential of dry forests under community management in Tigray, Ethiopia. Ecol Processes 6: 20.

Tadese S, Soromessa T, Bekele T, Bereta A, Temesgen F., 2019. Aboveground biomass estimation methods and challenges: A review. Int J Energy Technol Policy 9 (8):12-25.

Takimoto, A., P. K. R. Nair, and V. D. Nair. 2008. Carbon stock and sequestration potential of traditional and improved agroforestry systems in the West African Sahel. Agriculture Ecosystems & Environment 125:159-166.

Tesfay, H.M.; Negash, M.; Godbold, D.L.; Hager, H. Assessing Carbon Pools of Three Indigenous Agroforestry Systems in the Southeastern Rift-Valley Landscapes, Ethiopia. Sustainability **2022**, 14, 4716. https://doi.org/10.3390/su14084716.

Tetemke BA, Birhane E, Rannestad MM, Eid T., 2019. Allometric Models for Predicting Aboveground Biomass of Trees in the Dry Afromontane Forests of Northern Ethiopia. Forests 10: 1114.

Tura Bareke, Admassu Addi, 2021. Allometric equation for aboveground biomass estimation of *Galiniera saxifraga* (Hochst.) Bridson in Gesha-Sayilem forest, southwestern Ethiopia, Asian Journal Of Forestry Volume 5, Number 2, December 2021 E-ISSN: 2580-2844 Pages: 76-82 DOI: 10.13057/asianjfor/r050204

UNFCCC (United Nations Framework Convention on Climate Change), 1997. Clean development mechanism. http://unfccc.int/kyoto_protocol/mechanishm.

Walker SM, Murray, L, Tepe, T. 2015. Allometric Equation Evaluation Guidance Document –
Guidance Document for Lao PDR. Developed by Winrock International, on behalf Lao
PDR. Funded by JICA.

Wang, H., Hall, C.A.S., Scatena, F.N., Fetcher, N., Wu, W., 2003. Modeling the Spatial and
temporal variability in climate and primary productivity across the Luquillo Mountains,
Puerto Rico. Forest Ecology and Management 179, 69-94.

Wang, C., 2006. Biomass allometric equations for 10 co-occurring tree species in Chinese
temperate forests. Forest Ecology and Management 222: 9-16.

Wele, D. Ambachew, Singh, B, R., Lal, R. 2009. Soil carbon and nitrogen stocks under
chronosequence of cultivated and traditional agroforestry land use in Gambo District,
Southern Ethiopia. Soil Science Society of American Journal

Xu X, Medvigy D, Trugman AT, Guan K, Good SP, Rodriguez-Iturbe I. 2018. Tree cover shows
strong sensitivity to precipitation variability across the global tropics. Glob Ecol
Biogeogr. 27(4):450–60. (Xu X. et al 2018)

Yehualashet Belete, 2016. Species Specific Allometric Model For Biomass Estimation of Four
Selected Indigenous Trees In Egdu Forest, Welmera District, Oromia Region, Ethiopia
MSc thesis.

Yuen, J.Q., T. Fung, and A.D. Ziegler, 2016. Review of allometric equations for major land
covers in SE Asia: Uncertainty and implications for above-and below-ground carbon
estimates. Forest Ecology and Management, 360: p. 323-340.